DIG DEEP!

Bugs That Live Underground

Bumblebees

Omar Stone

PowerKiDS press

New York

Published in 2017 by The Rosen Publishing Group, Inc.
29 East 21st Street, New York, NY 10010

First Edition

Editor: Sarah Machajewski
Book Design: Mickey Harmon

Photo Credits: Cover (sky) Severe/Shutterstock.com; cover (background) ifong/Shutterstock.com; cover (bumblebee) Anatolich/Shutterstock.com; pp. 3–4, 6, 8, 10, 12, 14, 16, 18, 20, 22–24 (background) isaravut/Shutterstock.com; p. 5 https://en.wikipedia.org/wiki/Bumblebee#/media/File:Apoidea.jpg; p. 6 (map) BigBigbb1/Shutterstock.com; p. 7 DansPhotoArt on flickr/Moment Open/Getty Images; p. 9 (pollen) schnuddel/E+/Getty Images; p. 9 (bumblebee) Ivaschenko Roman/Shutterstock.com; p. 9 (background) amenic181/Shutterstock.com; p. 11 SMC Images/Oxford Scientific/Getty Images; p. 12 Michael Durham/Minden Pictures/Getty Images; p. 13 Ji-fang Zhang/Moment Open/Getty Images; p. 15 dwphotos/Shutterstock.com; p. 17 Photo Researchers/Science Source/Getty Images; p. 19 (inset) Bill Kennedy/Shutterstock.com; p. 19 (main) Jari Sokka/Shutterstock.com; p. 21 Roman023/Shutterstock.com; p. 22 Trofimov Denis/Shutterstock.com.

Library of Congress Cataloging-in-Publication Data

Names: Stone, Omar, author.
Title: Bumblebees / Omar Stone.
Description: New York : PowerKids Press, [2017] | Series: Dig deep! bugs that live underground | Includes index.
Identifiers: LCCN 2016004026 | ISBN 9781499420463 (pbk.) | ISBN 9781499420487 (library bound) | ISBN 9781499420470 (6 pack)
Subjects: LCSH: Bumblebees–Juvenile literature.
Classification: LCC QL568.A6 S796 2017 | DDC 595.79/9-dc23
LC record available at http://lccn.loc.gov/2016004026

Manufactured in the United States of America

CPSIA Compliance Information: Batch #BS16PK: For Further Information contact Rosen Publishing, New York, New York at 1-800-237-9932

Contents

Meet the Bumblebee

Which bug is black, yellow, and fuzzy? It's the bumblebee! Bumblebees aren't the only kind of bee in the world, and it's easy to confuse them with other kinds. One way to tell bees apart is to look at where they live. If their home is in, on, or near the ground, they could be bumblebees!

People usually see bumblebees when they're flying around our backyard or **pollinating** flowers. We can learn a lot about bumblebees just by watching them. We can also learn a lot from exploring their home. Let's dig deep to uncover more about this interesting creature.

Bumblebees are known for making nests in or near the ground.

Dig Deeper!

Female bumblebees have a stinger, and they can sting a person or animal more than once. However, they only do this when they feel threatened.

Where to Find Them

More than 250 **species** of bumblebees live around the world. They live throughout North America, South America, Asia, and Europe. Bumblebee species can be found in Africa, north of the Sahara. Bumblebees aren't native to Australia, but they have been introduced to the island of Tasmania.

Bumblebees prefer areas that are **humid** and cool. Flowers grow well in this type of **climate**. But they can live in many other kinds of **habitats**—they can even survive in the Arctic. Bumblebees don't build nests in areas that get a lot of sunlight because the nest gets too hot for the bees.

where bumblebees live

Predators that hunt bumblebees include birds, spiders, and wasps.

Looks Like a Bumblebee

People often confuse bumblebees with other bee and wasp species. The easiest way to identify a bumblebee is to look for a furry body. The yellow and black fur keeps bumblebees warm. This allows them to survive in colder areas and climates. The fur helps bees collect pollen from flowers. It also helps bees feel things around them.

Like most **insects**, bumblebees have two sets of wings and six legs. Two antennae, or feelers, allow a bumblebee to sense its surroundings. Bumblebees have compound eyes. That means each eye is made of many smaller parts that work together. Bumblebees also have three simple eyes that detect light and dark.

Bumblebees stand out from other bee species because of their furry body. Pollen sticks to their fur as bumblebees fly from flower to flower.

Dig Deeper!

A bumblebee will raise one of its middle legs when it feels threatened. This means, "Stay back or I'll sting you!"

Bumblebee Families

You will find three kinds of bumblebees inside a nest. All three kinds start life as eggs and then develop into **larvae** before growing into adults. Most of the bees in a nest are female. They're hard workers! In spring, it's the females' job to find food and build the nest.

Later in the summer, some eggs develop into males, and some develop into new queen bumblebees. After leaving the nest, the males **mate** with the queens and die soon after. Then the queens find an underground place to **hibernate** during the winter. In spring, the queens search for places to start new colonies.

The queen bumblebee is the largest bee in the nest. The queen is in the center of this photograph, and female worker bees surround it.

queen bee

Starting a New Colony

In areas that have four seasons, a bumblebee colony lasts from early spring until late summer. Colonies in warmer areas may last more than one year. When a queen wakes up in spring, it must find early flowers so it can gather pollen and nectar. This allows it to eat and prepare to lay eggs. For now, the queen hides near a source of food.

Soon the queen starts looking for a place to make a nest. It can be seen flying close to the ground at this time. Good places for a bumblebee nest include empty rodent holes, low-lying shrubs, and long grass, as well as sheds and garages.

queen bee

In spring, queen bumblebees can be found flying in zigzagging paths and crawling in and out of tiny holes. They're looking for a good place to start a nest!

Building the Nest

A bumblebee nest starts when the queen selects a place to settle down. She builds a mound of pollen and lays six to eight eggs on it. Then she covers the mound with a wax she makes in her body. It takes about four to five weeks for young bees to hatch and become adults.

These adult female bees take over for the queen so she can rest, lay more eggs, and care for larvae. The worker bees find sources of food, collect pollen and nectar, and bring them back to the nest. A finished nest can be as small as a tennis ball or as large as a shoebox.

Worker bees use their wax to build cells to hold eggs, pollen, and nectar. Building these wax structures allows the nest to grow.

Dig Deeper!

Most adult bumblebees live about four weeks.

Life in the Nest

Bumblebee nests are much smaller than the nests of other kinds of bees. Honeybee hives can hold around 50,000 bees! However, a bumblebee nest only holds up to 400 bees. Most of them are female worker bees.

Worker bees search for pollen and nectar out in the world, but there are also many jobs for workers inside the nest. For example, eggs and larvae need to be kept between 86 and 90 degrees Fahrenheit (30 and 32 degrees Celsius). The queen uses her body to keep the eggs warm. If the nest gets too hot, worker bees stand by the entrance and use their wings to fan cooler air into the nest.

Dig Deeper!

Special workers called housekeepers help keep the nest clean. These bees are either young bees or injured bees that can't fly.

Bumblebee eggs need lots of care if they're to survive. Luckily, there are plenty of bumblebees to get the job done!

It's All About the Pollen

Adult worker bees spend a lot of time searching for food. The colony couldn't survive if they didn't. A worker bumblebee lands on a flower blossom and crawls into it. It uses its long tongue to drink nectar. While there, its furry body gets covered with pollen. When the bumblebee moves on to the next flower, it brings pollen with it. This allows the second flower to create new flowers.

Bumblebees are pollinators. This means they aid people by spreading pollen from one flower to another, helping important crops to grow. Bumblebees have sacs for storing pollen on their back legs.

Dig Deeper!

A bumblebee's quickly moving wings help stir up the pollen in a flower, allowing more to stick to the bee's fur.

The sacs on a bumblebee's back legs are called pollen baskets.

Threats to Bumblebees

People need all kinds of bees, including bumblebees. Without them, the crops we depend on would disappear. However, bees face many threats. Many scientists think the world's bees are in danger of dying out.

Since the late 1990s, beekeepers have noted a decline in bee populations. There are several reasons for this. New building projects take habitats away from bumblebees, giving them fewer places to build nests. Modern **pesticides** aren't meant to harm bumblebees, but many of them do. Climate change is another danger threatening bee colonies around the world.

Beekeeping is a fun hobby! These backyard beekeepers have been trained to handle their bees without getting hurt.

You Can Help!

Much is being done to help bumblebees. You can help, too! You can read about bumblebees and teach others about them. Don't use pesticides in your garden unless you're sure they're safe for bees.

One of the coolest ways to help bumblebees is by creating a bee habitat. There are many plans for habitats on the Internet. You need to make your habitat look inviting to a wandering queen bumblebee. The nest should have small openings. It should be placed in a shady area so it doesn't get too hot. Don't forget to plant the right kinds of flowers to keep bumblebees healthy and happy.

Glossary

climate: The average weather conditions in an area over a period of time.

habitat: The natural home of a plant or animal.

hibernate: To spend winter in an inactive state.

humid: Marked by having a high water content in the air.

insect: A bug that has six legs and one or two sets of wings.

larva: A stage in an insect's life between the egg and the adult stages. The plural is "larvae."

mate: To come together to make babies.

pesticide: A chemical used to hurt insects or other organisms that harm plants.

pollinate: To bring pollen from one plant to another, allowing it to make new plants.

species: A group of living organisms that have similar traits and are able to mate.

Index

Websites

Due to the changing nature of Internet links, PowerKids Press has developed an online list of websites related to the subject of this book. This site is updated regularly. Please use this link to access the list: www.powerkidslinks.com/digd/bee